A Manual of Bamboo Hybridiza

A Manual of Bamboo Hybridization

INBAR Technical Report No. 21

Zhang Guangchu

Guangdon Forestry Research Institute, Guangzhou, China

Utrecht • Boston • Köln • Tokyo, 2002

VSP BV
P.O. Box 346
3700 AH Zeist
The Netherlands

Tel: +31 30 692 5790
Fax: +31 30 693 2081
vsppub@compuserve.com
www.vsppub.com

©VSP BV 2002

First published in 2002

ISBN 90-6764-356-4

Printed in The Netherlands by Ridderprint bv, Ridderkerk.

Contents

Foreword

All the major crop plants have been subjected to genetic improvement, either by selection and propagation or by breeding. Bamboos have received scant attention from plant breeders despite their importance as crop plants due their unpredictable and uncontrollable flowering habits and to a limited understanding of genetic variation amongst the existing species. The potential for new, improved bamboo hybrids is enormous. The demand for bamboo is increasing worldwide and the diversity of uses to which it is put is growing steadily. Increases in demand can be met by increasing the areas of bamboo plantations, but improvements in the quality of raw bamboo can only be met by selection and breeding.

Professor Zhang Guangchu of Guangdong Forestry Research Institute has worked on bamboo hybridization for almost thirty years and has amassed a wide range of skills and experience. She has produced hybrid bamboos that are now being grown commercially in South China. INBAR recently invited her to distil her experiences and make them available to a wider audience and this manual is the result. The manual refers primarily to the bamboos of southern China where the author is based, but the principles and techniques are applicable worldwide.

This manual is one of the products of INBARs Ecological Security programme, which aims to improve the genetic diversity, conservation and management of bamboo and rattan resources, and to promote their use in environmental protection and rehabilitation. We hope that it will be the catalyst for scientists, technicians, foresters, farmers and individuals to undertake bamboo hybridization in their own regions, to stimulate relevant research and to promote the wider acceptance and use of hybrid bamboos.

Dr. Lou Yiping
Programme Manager
Ecological Security

Dr. Ian Hunter
Director General

Acknowledgements

The main sources of reference for this book are the many papers produced from the Guangdong Forestry Research Institute (GDFRI). I am extremely grateful to those at GDFRI with whom I have had the pleasure of working: Mr. Chen Fushu, Mr. Yang Aiguo, Ms Wang Yuxia, Mr. Tan Yuanjie and Mr. Li Xingwei. This book is based on the results of much of their diligent work.

During the writing and editing of the book Mr. Andrew Benton made many useful suggestions and translated the book into English. I also wish to thank the International Network for Bamboo and Rattan who funded the production of the book.

Introduction

Bamboos grow rapidly, have a wide range of uses and play important roles in the rural economies of many tropical and subtropical regions. Those with durable culms and close grain are often used for scaffolding and are suitable as frames for buildings. Bamboo plyboard and bamboo flooring are good ornamental products in the modern home. Bamboo fibres are long and thin and can be used to make paper. Bamboo shoots have a distinctive flavour, are full of nutrients and are a popular natural food. The sap of bamboo leaves and tender shoots contains many physiologically active factors that can be used in medicine or in health foods. Many species are very graceful and are often planted as ornamentals in parks and gardens and some are used for environmental improvement. There is also much potential for developing bamboos as a source of energy in the future.

Experience of genetic improvement of forest trees has indicated that it frequently brings enormous economic benefits to producers. So, is it possible to genetically improve bamboos? The greatest obstacle to breeding bamboos is the length of the flowering cycle (the period of vegetative growth before flowering), which usually lasts from a few dozen to more than one hundred years, depending on the species. This evidently precludes the use of the conventional breeding methods used for other forest tree species. Bamboos can, however, be propagated vegetatively very easily. In most cases elite strains retain their genetic characteristics after vegetative propagation and are unlikely to loose them due to cultural conditions. On this basis a bamboo-breeding programme can be adopted that involves the sexual production of an F_1 generation by hand pollination of naturally flowering plants, followed by selection of elite offspring from the products of those crosses and then mass vegetative propagation for planting out and use.

Research into genetic improvement of bamboos commenced at Guangdong Forestry Research Institute (GDFRI) in 1972. Since then our research programmes have covered the following subjects: Biological characteristics of bamboo flowering and fruiting, determination of pollen viability, pollen storage techniques, hybridization techniques, cultivation of hybrids, selection and appraisal techniques, cytological investigations, rapid propagation of elite progeny by micropropagation and *in vitro* flowering. This book is based upon information accumulated during these many years' experience of working on bamboo breeding. It is a practical guide and a reference for others involved in breeding bamboos. It is expected that more comprehensive books on the subject will emerge as work on bamboo breeding increases. The breeding and raising of hybrids involves many branches of science, such as genetics, taxonomy, anatomy, physiology, cytology, cultivation and wood science. Many detailed books have been published on these subjects and the reader is recommended to consult standard reference works for more information if required.

1   Hybridization of bamboos

1.1. BIOLOGICAL CHARACTERISTICS OF FLOWERING AND FRUITING

1.1.1. Forecasting flowering

The most reliable indication that flowering is imminent is the production of small-bladed leaves on the branch tips of both young and old culms (Figs 1 and 2). The blades of these leaves become progressively smaller as one moves towards the apex of the branch. The branch subsequently stops growing and the tip swells as the flower bud is formed within it. The flower bud will then develop into a spikelet on which the flowers are borne. This may take from one to three and a half months depending on temperature. In summer buds forming at the beginning of a month may flower by the end of that month but those forming in the autumn will not flower until the following spring.

There are other indications that flowering is about to commence, but these are less immediately apparent. Prior to flowering the vigor of the plant is reduced. The seasons' new shoots emerge earlier than in non-flowering years and are small, less vigorous and significantly fewer in number. At the same time the carbohydrate content of the culm and rhizome increases and the nitrogen content, falls resulting in a high C : N ratio.

1.1.2. Flower structure and flowering

Spikelets and florets are the basic floral components of bamboos. Each spikelet is a collection of florets arranged successively along an axis. They are usually composed of between three and ten florets each. The florets can be divided into two types depending upon structure and flowering characteristics:

1) The open-flowering type. In this type the floret is composed of flowering glumes (palea and lemma), stigmas, stamens (anthers and filaments) and lodicules. When flowering the lodicules swell with water and force apart the palea and lemma. The anthers then emerge and the three plumose stigmas expand. This type of flower is easier to pollinate by hand because the male and female parts mature at the same time. The florets of *Bambusa* are of this type (Figs 3 and 4).

2) The closed-flowering type. In this case the floret lacks lodicules and so the glumes do not open. The stigmas emerge over 10 days before the anthers and have already dried out and died by the time the anthers emerge. In this type of flower manipulative pollination is much more difficult and careful observation is required to hand pollinate at the right time. The florets of *Dendrocalamus* are of this type (Fig. 5).

Flowers can be found on bamboos at all times of the year, but the main flowering period in South China is March to July. Sympodial bamboos usually flower earlier,

Figure 1. Normal leaves.

Figure 2. Small bladed leaves at the tip of a flowering branchlet.

from March to May, and monopodial and amphipodial bamboos flower from April to July. During flowering the flowers open in batches. The development of all the flower buds of each batch is concurrent and they all flower as a group, opening within 5–7 days, with a 2–3 day peak period. The second batch of flowers opens about 12–30 days later, depending upon the temperature. Lower temperatures delay opening and vice-versa. Some bamboos, such as *Dendrocalamus* species, have spikelets arranged in unlimited infloresences and often flower a second time in October–November.

Bamboos usually flower in the early morning. The anthers are extended and release pollen about one or two hours later. After approximately four to six hours the ovary is fertilized and the stigmas wither. Spikelets with unfertilized florets usually fall within 5–7 days. At the same time new spikelets are being formed from buds at the base of the old spikelets and these will flower later.

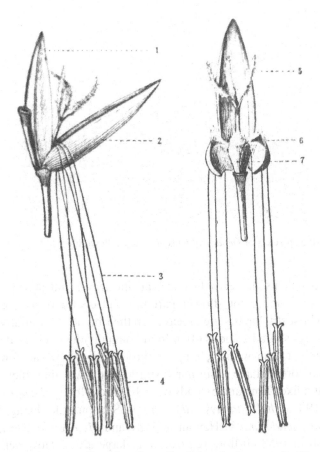

Figure 3. Structure of an open-flowering floret (1: Palea, 2: Lemma, 3: Filament, 4: Anther, 5: Stigma, 6: Lodicule, 7: Ovary).

Figure 4. Flowering spikelets of *Bambusa gibboides* (open-flowering type).

Figure 5. Flowering spikelets of *Dendrocalamus sp.* (closed flowering type).

Some bamboo plants flower only once, all the leaves fall and the whole plant dies afterwards. Some plants flower partially and continuously for many years before dying whilst still having green leaves on their branches. Initially flowering in bamboos is always sporadic and is often found on individual culms. It subsequently spreads gradually to the whole clump or throughout the whole forest. Usually natural bamboo forests have a similar history of renewal thus their flowering is concentrated at a fixed time and extends over the whole forest. *Dendrocalamus spp.* in Yunnan in 1981, *Phyllostachys pubescens* in Shaoping in Guangxi in 1963–4, *Schizostachyum pseudolima* in Hainan in 1981 and *Melocanna baccifera* in India and Bangladesh in 1959 all flowered over very large areas. However in cultivated forests individual plants or culms are often seen in flower, but population-wide flowering is very rare. This is especially so in forests of sympodial bamboos, such as *B. pervariabilis*, *B. textilis*, *D. giganteus* and *D. latiflorus*, when the clumps originate from different locations.

1.1.3. Pollen characteristics

The viability of bamboo pollen is related to flower quality and climatic factors when flowering. Viable pollen starts to germinate in 15–20 minutes on a suitable medium. The pollen tube extends longer than the diameter of the grain after 20–30 minutes and by 30–60 minutes the numbers of grains germinating has stabilized and a few of the pollen tube ends have split, indicating completion of the germination process. The pollen viability of some common South China bamboos is shown in Table 1. Bamboo pollen is thin walled and looses viability easily if it swells with water or dries out in direct sunlight or a dry atmosphere. Pollen life can be as much as 2–4 hours if the anthers are kept in a cool, shady place, but pollen will loose its viability in half an hour if no protection is provided.

Table 1.
Pollen germination of some bamboos of South China. Pollen germination in a medium containing 5–10% sucrose and 10 ppm boric acid

Species	Germination rate
Bambusa pervariabilis	2.9–14.8%.
Bambusa sinospinosa	4.3–14.3%
Bambusa textilis	3.4–7.2%
Dendrocalamus latiflorus	5.4–40.4%
Phyllostachys pubescens	26.5–64.1%

1.1.4. Seed setting

The rate of seed setting is closely related to pollen viability. If the pollen viability is high then the seed setting rate is also high and if it is low then the opposite is true. In the majority of bamboo species the natural seed setting rate is very low. Commonly 90% or more of flowers do not set seed. This is due to three factors:

1) Rapid loss of pollen viability.

2) Poor light conditions. In a bamboo forest, flower development is often suboptimal due to the shady conditions and results in high percentages of aborted pollen. This reduces the fertilization and fruiting rate. Transplanting a flowering forest bambooculm to a more open environment will increase the fruiting rate.

3) Damage by pests and diseases during the flowering period.

Once fertilized, the ovary swells rapidly. The formation of seeds takes 12–25 days in *Bambusa textilis*, *B. pervariabilis* and *B. sinospinosa*, 20–30 days in *Dendrocalamus latiflorus* and *D. minor* and 50–70 days in *Phyllostachys pubescens*. The cooler the temperature, the longer the seeds take to develop.

Once the seeds are ripe they fall very easily. Fresh seeds have about 50–60% viability but this declines as the storage time increases. After one year seed viability is completely lost. In order to raise seedlings the seeds should be sown as soon as they are harvested.

1.1.5. Causes of flowering

Bamboos gradually develop vegetatively and flower and seed when they reach sexual maturity. The length of this vegetative growth process (also called the intermast period or flowering interval) differs greatly between species. Flowering intervals vary from a few years through to many dozens of years (see Table 2).

The physiological mechanisms of bamboo flowering have still not been elucidated and many different theories have been proposed. Some believe the bamboos have an internal alarm clock that is set off at flowering time and all the culms of plants from the same origin then flower at the same time. Some think that a reduction in the growth of the vegetative cells creates an increased C:N ratio in the plant's tissues that then causes flowering. Some people think that external factors such as

Table 2.
Flowering intervals of some bamboo species

Species	Flowering interval (years)
Bambusa arundinacea	30–45
Bambusa tulda	30–60
Dendrocalamus hamiltonii	30–40
Dendrocalamus strictus	20–65
Melocanna baccifera	30–45
Ochlandra travancorica	7
Phyllostachys bambusoides	50–60
Phyllostachys nigra	40–50
Phyllostachys pubescens	50–60
Thyrsostachys oliveri	48–50

damage by pests and diseases, cutting, burning or drying out induces the growth of sexually-programmed cells and hence flowering. There are also protagonists of the cyclical aging/renewal theory. As Table 2 shows there are only a few bamboos with a relatively fixed intermast period. Apart from the genetic characteristics of bamboo flowering, environmental conditions and mans' influence may also have a specific delaying or promoting effect. Bamboos do not flower and seed immediately on reaching sexual maturity and the length of time in flower is often very variable.

1.2. HYBRIDIZATION TECHNIQUES

The stages of bamboo hybridization are as follows. See separate document called "Stages".

1.2.1. Selection of hybridizing parents

When considering the selection of parents for hybridization you should first clarify the breeding objectives. You may wish to increase timber yield and improve timber quality, to produce varieties with improved shoot flavour, increased tolerance to adverse environmental conditions or better resistance to pests and diseases. Selection of parents with good characteristics relevant to the breeding objectives will produce good hybrids. Random selection and hybridization of parents is a waste of time and energy. In fact it may ultimately be impossible to select a useable hybrid from the progeny of such crosses.

Selection of parents includes two stages: selection of the hybrid parents, and selection of culms from the chosen parents. The main principles in selecting the parents themselves are outlined below.

1.2.1.1. Selection based upon the main objectives of the breeding programme.
Each species has a specific suite of different characteristics that make it suitable for

Stages of hybridization

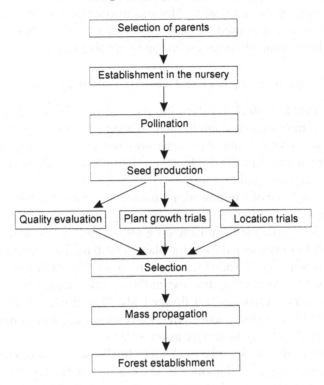

Scheme 1. Stages of hybridization.

particular uses. In order to breed a new bamboo hybrid with broad culms and good quality wood which could be used as a substitute for *P. pubescens* but which would not compete with *Cunninghamia lanceolata* for land, we selected the broad-culmed *D. latiflorus* and the high timber-quality *B. pervariabilis* as parents. In areas of high soil erosion which need rapid revegetation the aim may be to breed strongly resilient bamboos and here it would be worth considering the drought-resistant *D. strictus* and the cold-resistant *B. multiplex*.

1.2.1.2. Selection of parents with complimentary characteristics. Selecting parents such that each has some of the desired characteristics, and together they have all, will ensure that some of the hybrids have the ideal combination of characteristics. For example crossing *B. sinospinosa* (large with hard, dense wood and very suitable for pulping but tough thorns at the base of the culms make harvesting difficult) with *B. textilis* (excellent source of fibres for pulping, basal nodes of the culm are branchless but culms are narrow and productivity is low) should produce some hybrids with broad, thornless culms which are suitable for pulping.

Most of the F_1 generation have characteristics interposed between those of each parent but in practice it is obvious that the influence of the maternal parent is greater than that of the paternal parent. After selecting the plants to hybridize decide which has the better suite of characteristics. This can then be used as the maternal parent and the poorer of the two used as the pollen source. This will ensure a greater chance of the better characteristics appearing in the offspring.

1.2.2. Transplanting breeding stock and establishment of a parent collection

Transplanting breeding stock involves the removal of a portion of the flowering culms from a clump and their establishment in a nursery or in pots. Add potassium and phosphorous fertilizers and due to the improvement in the nutrition and light conditions normal growth and development of the flower buds will be promoted and the seed setting rate will increase.

The majority of bamboos in flowering nurseries have been collected from naturally occurring sporadic flowerings. The best way of finding a sporadically flowering bamboo is to visit the main bamboo growing areas in the autumn, after the seasons growth has ceased and look for plants that have just started flowering or that have culms with small-bladed leaves at the tips of the branches. Choose 1–2 year old culms for transplanting because at this age the transplants survive easily and will produce new branches and flower buds. The flowers will also be good quality. The bud germination and rooting percentages of older bamboo culms is poor and they are less likely to survive as transplants.

The culms selected should not be too long or too broad — a diameter of 4–8 cm is ideal. Cut off the top of the culm to leave a propagule 1.5–2.0 m long and be careful not to damage the rhizome when removing the culm from the clump. Propagules much larger than this are less likely to survive. In order to prevent water loss from the culms during transport it is best to wrap them in plastic but make some ventilation holes and be sure to transport the culm rapidly to the nursery and plant it immediately.

The nursery should be sited in a well-lit spot in which it is convenient to work but it does not need to be isolated from other bamboos as the chances of cross-contamination are very low indeed. Plant at 2–3 m spacings and at an angle to make hybridization manipulations more convenient. Alternatively the tips can be cut off to make the material shorter. Potassium and phosphorous fertilizers should be worked into the bottoms of the planting holes. Otherwise the methods are the same as normal bamboo forest establishment methods.

1.2.3. Season of hybridization

Bamboos mostly flower from February to June and seed production is greatest in May and June. Seeds produced from earlier flowerings are generally of better quality because the temperature during ripening is lower and ripening takes longer. This produces well-filled grains with high germination rates and high rates of plantlet formation. As the temperature increases seeds ripen faster and their quality declines. Above 30°C the supply of nutrients to the developing seeds is poor, resulting in seeds that are not fully filled and a greater proportion of albino seedlings. Thus best results are obtained by hybridizing during the early part of the main flowering period.

A few bamboo species such as *P. pubescens* and *D. latiflorus* flower a second time in late autumn (November–December) and although there are fewer flowers the seed formation rate is still high and cross-pollination is usually successful.

1.2.4. Regulation of flowering period

Bamboo pollen looses viability very rapidly and a supply of fresh pollen is vital for hybridization. This is only possible when the flowering of both parents coincide. In order to obtain a continuous supply of fresh pollen, transplant the parent culms at different times in order to stagger the flowering times. For example if flowering culms are transplanted in September, November, January, March and May then by the time the culms transplanted in September have finished flowering, those transplanted in November or even January will be starting to flower and thus there will be a supply of fresh pollen each day throughout the hybridization period.

1.2.4.1. Pollen storage. It is possible to store pollen for short periods of time. The method is as follows. Place freshly collected, unopened anthers in a small test tube filling it about one third full. Bung the top with a moist bung of cotton wool to ensure they do not dehisce whilst in storage (Fig. 6). The bottle is then placed in the fridge for storage at a temperature of about 4 °C. Note that the bung should not be dripping wet. Using this method to store pollen of *P. pubescens* the germination rate was 50.7% when fresh, 28.3% after 5 days and 9.6% after 7 days but it was still effective for hybridization (see Table 3).

1.2.5. Testing pollen viability

Testing pollen viability before pollinating avoids ineffective pollination and helps in the analysis of the compatibility of the parents. There are many methods for checking pollen viability. The most reliable method involves sprinkling fresh pollen on a medium suitable for germination and then calculating the germination percentage. We have developed a simple and convenient method of testing pollen viability using the enclosed hanging drop method as a basis, with modifications to account for the particular characteristics of bamboo pollen (see below).

1.2.5.1. Components of medium.

Cane sugar (sucrose)	10 g
0.1% sodium borate solution	1 ml
Make up with water to	100 ml

Note: The medium must only be used fresh. It does not need to be sterilized but after the first use the surplus cannot be stored for later use.

Figure 6. Pollen storage.

Table 3.
Loss of germinability of *Phyllostachys pubescens* pollen during storage

Date (1974)	Storage time	Germination (%)	Loss of germination (%)
June 23	Fresh	50.7	0
June 24	1 day	46.3	8.6
June 26	3 days	35.5	30.0
June 28	5 days	28.3	44
June 30	7 days	9.6	80

1.2.5.2. Tools.
A piece of wood 6 cm × 2.2 cm × 0.6 cm deep in which two circular holes, each 1.5 cm in diameter, have been drilled side-by-side.
Glass cover slips
Slides
3–4 culture dishes
2 × 60 ml dropping bottles
2 × 250 ml beakers
1–3 small test tubes
One pair of fine tweezers
One bamboo spatula (with one end flattened)
Microscope

1.2.5.3. Materials.   Fresh bamboo pollen.

1.2.5.4. Method.

1) Soak the wood block in clean water until saturated.
2) Remove the block from the water and place it on the top of a slide ensuring that the slide completely covers the two holes in the wood.

3) Take a clean coverslip and place a drop of germination medium on the surface. Sprinkle on some pollen then rapidly turn the coverslip over and place it face down over one of the two holes in the wood.

4) Check that the wood is in full contact with both slide and coverslip. If there are any air bubbles place a drop of distilled water onto the intersection of wood and coverslip or slide to close the gap. Be very careful not to break the drop. In this way the pollen germinates enclosed in a humid environment.

5) Replicate four times for each plant.

6) Count the grains one hour after inoculation. For each pollen sample count four fields of view. Germinated grains are those in which the pollen tube is longer than the diameter of the grain. Record the total number of grains per field of view and the number of germinated grains.

7) The germination percentage is calculated as (number of grains germinated in 16 fields of view / total number of grains in 16 fields of view) × 100.

Note the particular characteristics of this technique are: 1) there is no need to sterilize the medium; 2) a piece of wood is substituted for a glass ring and water is used as the sealant; 3) using only fresh medium prevents changes in sucrose content of the medium due to fermentation which might influence the germination of the pollen.

1.2.6. Emasculation

Bamboo flowers are hermaphrodite and wind pollinated. Although the structure of the flowers prevents self-pollination there is the possibility of pollen transfer between flowers on the same culm. The potential exists for some seeds to be formed by self-crossing, so to avoid mixing up seeds produced by hand pollination and natural pollination it is best to remove the stamens from all the flowers of the maternal parent. Many of the methods of emasculation used in breeding cereals would be appropriate, but in practice the most convenient method is to use a pair of tweezers to pluck out the stamens from the flowers when they are opening in the early morning.

1.2.7. Selection of flowers

It is not possible to pollinate all the flowers produced during the main flowering period and it is advisable to thin the flowers at this time to reduce the unwanted consumption of nutrients. Those in the middle of the spikelet are healthy and robust and under natural conditions most of seeds are formed from them. The flowers at either end of the spikelet are small and less well developed so these can be thinned out using a small pair of scissors.

1.2.8. Pollination

Most flowers open from six to nine o'clock in the morning when the weather is shady and cool and the relative humidity is quite high. This is also the best time to hand pollinate.

1) Take fresh anthers from the paternal parent and spread them out on a clean, dry piece of paper. Keep them in a shady place and after about half an hour they will dehisce. Anthers stored in a test tube should be taken out and treated in the same way.
2) Use a fine-haired Chinese writing brush or a bamboo spatula to pick up some of the pollen and sprinkle a few grains on the stigma.
3) Mark the pollinated flower with a drop of paint of an eye-catching colour and hang a small cardboard tag on the node recording the parents, the number of flowers pollinated at that node and the date (Fig. 7).

The pollinated flower does not need to be bagged because the pollen looses its viability very rapidly and the natural pollination rate is very low such that there is no risk of contamination.

1.2.9. Post-pollination management

Shade and protection from rain should be provided between pollination and flower closing. If hybridizing in flowerpots move the pots to a shade house for cross-pollination and then place them outside in a sunny spot once the flower has closed. This will greatly improve the success rate.

In South China, bamboo flowers are often damaged by a presently unidentified species of insect. This species collects around the stigma sucking the sap, and then lays its eggs in the flower. After the flower has closed the eggs are enclosed within the glumes and the larvae hatch and eat the ovary tissue. This species damages flowers in all the flowering periods and damage can be as much as 90% or higher so

Figure 7. Maternal parent (*Bambusa gibboides*) after cross-pollination in the nursery.

vigilance is required. Two or three sprays of 0.1% DDVP or other similar insecticide will provide satisfactory control.

1.3. SEED COLLECTION AND NURSERY MANAGEMENT

Check to see whether fertilization has taken place seven to ten days after pollination. Hold the floret up to the light and let the light shine through the glumes to see if it contains a light green swelling the size of a grain of rice. This shows that the ovary has been fertilized and will probably develop into a seed. Make a record of it and check it regularly because about 2–3 days before the seed ripens a translucent paper bag should be placed over it to prevent loss should the seed fall once ripe. However it is not a good idea to bag too early as this may affect normal seed development. After collecting the seed, record its parents, date of hybridization, date of harvest, and the number of seeds collected from that culm.

Seeds should be sown immediately after collection. Sow singly in sterilized soil mixed with subsoil and one to two percent calcium superphosphate. Do not cover too deeply and keep moist before emergence, but avoid overwatering as the seed will rot. Germination takes place in 5–10 days. Once the shoots are 10–15 cm long they can be transplanted to pots. Spray the seedlings with a dilute solution of urea or other liquid nitrogen fertiliser every 2 weeks and the plants should grow to 1 m tall in the year of sowing. Plantlets can be planted in the experimental nursery the following spring at the same spacing between rows as in a normal bamboo forest. Other crops can be interplanted in the year of planting, but be careful to avoid damaging the bamboos.

1.4. CROSS-COMPATIBILITY OF PARENTS

Cross-compatibility is an indication of the relationship between bamboo species. High compatibility indicates a close relationship and low compatibility indicates a more distant relationship. From the point of view of breeding, the hybridization of closely-related parents produces normal seedlings in the F_1 generation, from which it is easy to select elite hybrids. In crosses between more distantly-related parents, many of the seedling generation do not grow well and it is difficult to select elite individuals from amongst them. For example, at GDFRI we conducted over 12,000 separate crosses between *B. pervariabilis* and *P. pubescens* and obtained only 34 seeds, none of which produced a seedling worthy of selection.

The results of over twenty crosses between seven species in four genera indicates the following rules:

1) The cross-compatibility of species of the same genus is quite high. E.g. *B. textilis* × *B. pervariabilis* had a fruiting rate of 13.6%, *B. textilis* × *B. sinospinosa* had a fruiting rate of 10.8% whilst *D. minor* × *D. latiflorus* had a rate of 22%.

2) Species in different genera but which grow in similar ecological conditions may or may not have relatively strong cross-compatibility. For example *D. latiflorus* as pollen source × *B. pervariabilis* or *B. textilis* gave 8.1–14.5% fruiting, but with *B. sinospinosa* the fruiting rate was only 0.6–1.6%.

3) Cross-compatibility of species from different genera that also differ in ecological requirements is low. E.g. *P. pubescens* as the pollen source × *B. pervariabilis* had only 1.3–3.8% fruiting rate, with *B. textilis* it was only 1.0–2.0% and with *B. sinospinosa* it was only 0.47–1.56%.

The low cross-compatibility of wide crosses like those above can be overcome by using mentor pollen from a more closely-related species. For example, in one experiment the fruiting rate in *B. pervariabilis* × *P. pubescens* crosses was 1.3–3.8%, but when *B. pervariabilis* was crossed with a mixture of pollen from *P. pubescens* and *B. textilis* the rate increased to 8.0%. The fruiting rate of *B. sinospinosa* × *P. pubescens* was 0.47–1.56%, but when a mixture of *P. pubescens* and *D. minor* pollen was used, the rate increased to 3.7%.

1.5. APPRAISAL AND SELECTION OF HYBRIDS

1.5.1. Selection of culm-use bamboos

Selection criteria should be based on breeding objectives. In order to raise construction-use bamboos particular attention should be paid to wood strength and durability. In order to raise bamboos for paper pulp then fibre type and content need to be taken into consideration.

The selection procedures are as follows:

1. Select elite clones from amongst the hybrids.

2. Mass propagate the elites and trial them in many different locations.

3. Subject the elites to anatomical and durability tests and select the best.

4. Subject these selected hybrids to further timber-quality or pulping studies and select the very best.

We have selected some excellent quality hybrids from our crosses using the above method and these have been distributed to propagation units.

1) ChengMaQing 1 (Fig. 8). Culms straight 13–15 m tall, 6–9 cm in diameter. Vigorous producer of shoots (6–18 new culms each year per clump). Timber quality good (see Tables 4 and 5). Culm quality, as tested by anatomical and physical methods, was found to be better than *D. latiflorus*, very similar to *B. pervariabilis* and marginally inferior to *P. pubescens*. In outdoor trials, culms have been found not to crack or rot easily and can be used as building frameworks and banana supports. The shoots are edible. Processed dried shoots are golden yellow in colour with a sweet flavour. The culms are alternately striped with many yellow-green stripes, and the plant can be used as an ornamental.

Figure 8. Hybrid ChengMaQing 1.

2) ChengMa 7. This was bred from *B. pervariabilis* as the maternal parent and *D. latiflorus* as the pollen source. It grows rapidly, is very adaptable, the culms are straight, and it has high productivity. The anatomical and physical characteristics of ChengMa 7 and its parents are shown in Tables 6 and 7.

The mechanical strength of ChengMa 7 is very similar to that of *B. pervariabilis* and it is considerably stronger than *D. latiflorus*. The total amino acid content of the shoots is 21.57 g per one hundred grammes, 7.05 g of which are essential amino acids. This is significantly higher than the widely grown *D. latiflorus* (16.07 g and 5.62 g respectively) and *B. beecheyana* var *pubescens* (16.87 g and 6.23 g respectively) and it can be used for both shoots and paper. Annual yields per hectare of 6–15 tonnes of fresh shoots and 2000 culms, 6–8 cm in diameter have been achieved.

3) QingMa 11. Culms straight, tall and graceful; branches emerging from upper nodes on the culm, clumps very attractive. This is an excellent ornamental hybrid. Fibres comprise 44.8% of the culm, mean fibre length is 2.3 mm and compression strength parallel to the grain is 3198 kg/cm^2. It is an excellent

Table 4.
Anatomical values of hybrid ChengMaQing 1 and it's parents

Species	Relative weight of fibres	Fibre length (L) (μ)	Fibre width (D) (μ)	L/D ratio	Fibre wall thickness (μ)	Vascular bundle density (per cm^2)	Specific density (g/cm^3)
ChengMaQing 1	43.8	1835	15.0	121.9	9.8	2.9	0.57
B. pervariabilis	42.5	1778	17.0	104.1	8.6	3.6	0.62
D. latiflorus	33.5	1530		90.0	9.2	2.4	0.47

Table 5.
Mechanical strength of hybrid ChengMaQing 1 and it's parents

Species	Compression strength parallel to the grain (kg/cm^2)	Tensile strength parallel to the grain (kg/cm^2)	Transverse bending stress (kg/cm^2)
ChengMaQing 1	553	1891	1421
B. pervariabilis	565	2025	1436
D. latiflorus	640	1868	1650

Table 6.
Characteristics of three year old wood of hybrid ChengMa 7 and it's parents

Species	Specific density (g/cm)	Vascular bundle density (per mm^2)	Fibre relative weight (%)	Fibre length/width ratio
ChengMa 7	0.55		41.9	95.0
B. pervariabilis	0.60	3.5	40.5	106.4
D. latiflorus	0.35	2.3	36.5	105.0

Table 7.
Characteristics of three year old wood of hybrid ChengMa 7 and it's parents

Species	Compression strength parallel to the grain (kg/cm^2)	Tensile strength parallel to the grain (kg/cm^2)	Specific density (kg/cm^2)
ChengMa 7	610	2361	1968
B. pervariabilis	692	3234	2147
D. latiflorus	454	1961	1262

variety for papermaking and weaving and is also cold tolerant; it does not suffer damage at $-3\,^\circ$C.

1.5.2. Early selection of culm-use hybrids

Early selection of elites is vital as plants take seven or eight years to become commercially productive. The basic factors that influence mechanical strength are found in the internal structure of the timber. Research at GDFRI has determined that there is a significant, or highly significant, relationship between mechanical strength, vascular bundle density and fibre relative weight. Thus the quality of the wood can be determined indirectly by measuring specific volume or cutting anatomical sections.

We have conducted comparative analyses on anatomical samples of nine species under the age of six years and on their mechanical strength when over six years of age. We regarded as elites bamboos with high vascular bundle density and high fibre relative weights. Those with high vascular bundle densities and low fibre relative weights or vice-versa were regarded as mid-quality species, whilst those with low vascular bundle densities and low fibre relative weights were regarded as low quality. The relationship between the vascular bundle densities and fibre relative weights of young, tender bamboos and the mechanical strength of the culm at age over six years old was positive and significant at the 0.1% level.

This indicates that anatomical evaluation at a very young age can help determine the nature of the wood some years later and the method is about 90% reliable. Generally this anatomical determination can be conducted from two years after sowing. Testing of mechanical strength must wait until the plant is mature, so using this method the time needed for evaluation can be reduced by at least 4–6 years, and the efficiency of breeding work can be increased.

1.6. SELECTION OF ELITE HYBRIDS FOR SHOOTS

Bamboo shoots are fresh and tasty, and rarely suffer from contamination by agro-chemicals. Moreover they contain a comprehensive complement of nutrients and are one of the superior vegetables. Following changes in eating habits the market requirement for bamboo shoots has been increasing. We have conducted research on genetic improvement of shoot-used bamboos with the aim of producing adaptable bamboo hybrids with tasty shoots, a high nutrition content and excellent processing properties. The research has included: collective evaluation of presently-available bamboo shoot resources, rapid propagation of elite cultivars *in vitro*, hybridization of local bamboos with introduced species, and improvements in the genetic quality of shoot-used bamboos. Rapid propagation of bamboos by tissue culture is discussed in the next chapter, but some of the methods for selecting shoot-use bamboos are outlined below.

1.6.1. Appraisal of bamboo-shoot species

The rationale for this programme is the selection of elite strains for propagation and distribution from amongst the presently available bamboo resources, but it is equally applicable to new hybrids. The appraisal procedure includes evaluating taste, productivity, shoot weight, edible percentage, nutritional value and colour of the processed product.

1.6.1.1. Taste.   This involves investigating flavour and texture:
1) Flavour.

1. Bamboos with shoots with a fresh and sweet flavour with no bitterness and a good fragrance are the best species.
2. Those with shoots lacking bitterness, but also lacking a fresh taste or fragrance are moderate species.
3. Shoots that are slightly or very bitter are secondary quality products.

2) Texture.
   The best quality fresh shoots should be really crunchy. Those that are soft are secondary quality products.

1.6.1.2. Yield.   The productivity of species with optimum quality shoots is relatively unimportant but if the shoots are grown for processing then higher yields are preferred.

1.6.1.3. Individual shoot weight.   Species producing shoots of 0.5–1.0 kg each in weight are ideal.

1.6.1.4. Edible percentage.   This indicates the percentage by weight of edible portion of the shoot, which is what remains once the sheaths have been removed. The higher the percentage the better the species.

1.6.1.5. Nutrition content.   Fats and carbohydrates are plentiful in the modern diet so only total amino acid and essential amino acid contents need consideration.

1.6.1.6. Colour after processing.   Shoots that are white after boiling are less bitter and may be used fresh. If they are slightly yellow then the shoots are somewhat bitter and are better canned.

1.6.1.7. Colour after canning.   Products which are a lustrous milky white or light yellow are good. If the shoots are slightly red and lack any lustre then the variety is unsuitable for processing.

1.6.1.8. Colour of processed dried shoots.   Products with a golden yellow colour and lustre are good. Poor quality is indicated by dark yellow, dark tan or even black colourations, mainly because consumers erroneously believe the product has altered during processing, and this affects marketability.

On applying the above standards to a comprehensive study of 20, mostly sympodial shoot-use bamboos we were able to categorize the vast majority into one of two groups: Sweet shoot bamboos and Bitter shoot bamboos.

Sweet shoot bamboos have mid-sized shoots with no bitter taste or only a hint of bitterness and do not need to be soaked for long before use. Because of this they loose only a small amount of nutrition, have a relatively fresh and tasty flavour and can be used fresh. If processed into dried shoots or canned shoots they have an excellent flavour. The only drawback is that the dried shoots are slightly dark in colour and those in cans have a slight red tint and are less lustrous. In this group of shoots we include *B. gibboides*, *B. oldhamii*, *D. brandisii*, *D. hamiltonii* and *D. semiscandens*. *D. hamiltonii* has the best-flavoured shoots and their protein, amino acid and essential amino acid contents are very high (see Tables 8 and 9). It is an excellent shoot-use species.

Bitter shoots contain differing levels of bitterness which must be removed by boiling or steeping for a few hours prior to use. This long period of boiling greatly reduces the nutrition content of the shoots. These shoots are of relatively poor quality but when processed they have a relatively good colour. They can be graded by considering productivity, flavour and colour of the processed productas the main criteria. *B. beecheyana* var. *pubescens*, *B. vario-striata*, *D. latiflorus*, *D. minor* and *P. pubescens* all fall within this category.

During our studies we also unexpectedly discovered that the shoots of *B. pervariabilis* were tasty and their protein, total amino acid and essential amino acid contents were higher than all the other species in the test. *B. pervariabilis* is thus a very good double-use bamboo (timber and shoots) and moreover in terms of hybridization it is a good parent. Crossing it with shoot-used bamboos produced hybrids with increased the nutrition content of the shoots and crossing it with timber-used bamboos improved the culm shape and timber quality of the offspring.

1.6.2. Selection of hybrids

We chose MaBan 1 (a hybrid produced from *D. latiflorus* × *D. hamiltonii*) for rapid propagation and promulgation after evaluating the projeny of hybrids between *D. latiflorus* × *D. hamiltonii*, *D. latiflorus* × *P. pubescens* and *D. latiflorus* × *D. minor* using the criteria described above.

1.6.2.1. Maban 1 (Fig. 9). Grows vigorously and produces many shoots. Shoots taste fresh, sweet and crispy, but are also tender with an excellent flavour. Shoot quality is significantly better than it's maternal parent, *D. latiflorus*, and its shoot productivity is higher than its paternal parent, *D. hamiltonii*. The total amino acid content of the shoots is 16.87 g per 100 g dry weight, and the essential amino acid content is 6.58 g per 100 g dry weight. This exceeds that of the widely grown shoot-used species *B. beecheyana* var *pubescens*, *B. oldhamii*, *B. vario-striata* and *D. latiflorus*. It is an excellent quality hybrid. The colour of the processed shoots lies between those of sweet and bitter shoots, and is better than shoots of *D. brandisii*

Table 8.
Nutrition contents of shoots of important shoot-use bamboos

Species	Protein (g/100 g dry weight)	Fat (g/100 g dry weight)	Total carbohydrate (g/100 g dry weight)
Bambusa gibboides	16.3	1.18	38.5
Bambusa pervariabilis	27.5	1.82	30.2
Bambusa vario-strata	21.0	71.19	38.5
D. latiflorus × *D. hamiltonii* hybrid no. 1	25.1	1.66	35.8
D. minor × *D. latiflorus* hybrid no. 5	24.2	1.67	37.7
Dendrocalamopsis beecheyana var pubescens	16.9	1.17	49.2
Dendrocalamopsis oldhami	20.4	1.09	35.7
Dendrocalamus brandisii	24.8	1.31	36.0
Dendrocalamus hamiltonii	23.5	1.41	35.8
Dendrocalamus latiflorus	22.1	1.34	35.4
Dendrocalamus minor	22.2	1.37	30.8
Dendrocalamus semiscandens	18.6	1.10	37.6
G. levis × *D. latiflorus*	21.3	1.40	32.4
Gigantochloa levis	16.6	1.18	41.0

Table 9.
Amino acid contents of shoots of some shoot-use bamboos

Species	Total amino acids	Essential amino acids
Bambusa gibboides	11.46	4.56
Bambusa pervariabilis	20.40	7.32
Bambusa vario-strata	16.08	5.97
D. latiflorus × *D. hamiltonii* hybrid no. 1	16.87	6.58
D. minor × *D. latiflorus* hybrid no. 5	15.69	5.55
Dendrocalamopsis beecheyana var pubescens	11.74	4.20
Dendrocalamopsis oldhami	14.07	5.42
Dendrocalamus brandisii	17.58	6.34
Dendrocalamus hamiltonii	17.70	6.55
Dendrocalamus latiflorus	16.35	6.00
Dendrocalamus minor	17.05	6.48
Dendrocalamus semiscandens	13.34	4.69
G. levis × *D. latiflorus*	14.38	5.35
Gigantochloa levis	11.93	4.29

and *D. hamiltonii*, and marginally inferior to *D. latiflorus*. This variety is suitable for peri-urban cultivation and the only drawback is that it is not porticularly drought resistant and is therefore only suited to moist environments.

Figure 9. Hybrid MaBan 1.

1.7. CYTOLOGY OF PARENTS AND HYBRIDS

Chromosomes are the carriers of genetic material. Studying bamboo chromosomes can help us understand the relationships between species and can help explain the genotype of the species or hybrid. Such studies are fundamental to understanding taxonomy and breeding. Counting bamboo chromosome numbers can help determine whether a plant is a hybrid of the expected parents, which is especially useful when using mentor pollen.

1.7.1. Method of counting chromosomes

Apical or intercalary meristems are used to investigate chromosomes. Root tips are ideal. In order to obtain fresh root tips wrap one-year-old secondary branches in a moist towel and place them in a temperature of 25 °C to encourage the roots to grow from the rhizome. Root tips can also be collected after rainfall from new culms in the field that are still undergoing extension growth.

There are many methods of preparing slides for counting chromosomes (see botany textbooks) but because bamboos have many chromosomes, and they are difficult to separate, they are often difficult to count. After trying many different methods I believe professor Li MaoXue's (Chen and Song, 1982) method is the most suitable for bamboo chromosomes. With this method the chromosomes separate well and are easy to count. The method is shown below:

1. Excise a root tip and place in a mixture of 1/2000 colchicine and 0.002 M of 8-hydroxyquinoline (8-quinolinol) for 3–5 hours.

2. Place in 0.75 M KCl solution for 15–20 minutes.

3. Transfer to a 3% solution of 1 : 1 cellulase : pectinase to break the cell wall.

4. Wash once or twice with distilled water and then place in water for 2–4 minutes.

5. Pour off the water and use a needle to break up the root tip.

6. Pour on a 1 : 3 mixture of methanol : glacial acetic acid, and steep for 20–30 minutes.

Table 10.
Chromosome numbers of some bamboos of South China

Species	Chromosome number
Bambusa affinis	70
Bambusa chungii	72, 64
Bambusa dissemulator var *albinodia*	64
Bambusa lapida	64, 52
Bambusa pervarabilis	72, 64, 56
Bambusa rutila	64
Bambusa sinospinosa	64
Bambusa sp	64
Bambusa textilis	72, 64, 56
Bambusa vario-striatus	96, 84
Dendrocalamopsis biciatricatus	72, 64
Dedrocalamopsis stenoauritus	68
Dendrocalamus latiflorus	72, 64, 48
Dendrocalamus minor	72
Hybrid *ChengMaQing 1*	68
Hybrid *QingMaCheng 14*	68

7. Remove the sediment, suck up a few drops of cytoplasm (cell sap) and place on a pre-frozen glass slide.

8. Place the slide in Geisma staining dye solution at pH 7.2 to dye for 30 minutes.

9. Take out the slide, wash and dry it gently, and then it can be observed.

Under the microscope select cells with well-separated chromosomes for counting and/or photography. For each bamboo hybrid prepare two slides and count the chromosomes in at least 30 cells of each.

The chromosome numbers of some of the bamboos of the southern subtropical part of China are shown in Table 10.

We have used the technique to investigate the cytology of a number of the hybrids we have produced at GDFRI. For example hybrid ChengMaQing 1, which is a cross between *B. pervariabilis* and *D. latiflorus,* has a chromosome number of $2n = 68$, which is coincidentally the sum of the two haploid numbers of the parents (*B. pervariabilis* is $2n = 64$, *D. latiflorus* is $2n = 72$). *B. textilis* pollen had a mentor role in the cross, and was not involved in fertilization.

2 Rapid propagation of elite hybrids

The traditional method of propagating bamboos is by divisions from the mother clump (offsets). Healthy one year-old culms with rhizome and roots attached are separated from the clump and then tipped-back to about 1–1.5 m. They are then transplanted to their new location. However, with this method only one new plant can be produced per culm. This is important when attempting to mass propagate a new hybrid. The propagules are very large and difficult to transport and the survival rate is low. It is also very difficult to supply large numbers of propagules for planting large areas.

Propagation of bamboos at the seedling or nursery stage produces many new plants from each culm. They grow vigorously, have well-developed root systems and a high survival percentage when transplanted to the field. The plants thus formed establish rapidly and produce many shoots. Some of the commonly used methods of propagating sympodial bamboos at the seedling or nursery stage are outlined in this section.

2.1. NODAL CUTTINGS OF SEEDLINGS

1. Select healthy, normal 2–3 year old culms, sever them close to the ground and remove the apical portion less than 1.5–2 cm in diameter.
2. Remove all the branches at each node except for one thick branch and then tip these back to 2 cm above their first node.
3. Cut the culm into one- or two-noded propagules.
4. Plant the propagules horizontally at 15 × 40 cm spacings in a well-prepared seedling bed with the branches pointing out to either side.
5. Cover with soil to a depth of 3–5 cm, mulch with a layer of straw (or other porous mulching material) and water thoroughly.

Bud germination activity starts within 5–7 days of planting and roots are formed within 40–100 days. Success rates of about 50% can be expected and may be as high as 80% or more.

The success rate is affected by the following factors.

1. Culm age. Many years of research have shown that two year-old culms have the highest success rates. Although one year-old bamboo culms are the most vigorous, they contain few stored nutrients and have limited resistance to environmental stresses. They require intensive management and will die easily if neglected. Two-year old culms grow vigorously, have mature tissues and contain many stored nutrients. The success rate is high and the plantlets are healthy. The tissues of three- and four-year old bamboo culms have already started to grow old, their buds do not germinate readily and the growth of the propagules is poor.

2. Time of year. Bamboo cuttings should be taken during the dormant period or just as the buds start to shoot in the early spring. In Guangzhou the ideal time is from the end of February to the end of March. If taken any earlier, the weather is too cold for the cuttings to germinate in the soil and they will gradually loose their viability. If taken too late when the temperature is higher, evaporation is greater and the plant from which the cutting was taken will have already produced branches and leaves, which reduces the quantity of stored nutrients in the culms used for propagation. Our studies on *B. chungii* revealed that at the beginning of March the survival rate of cuttings was 80% and at the end of March it was 55.3–63.6%. Cuttings taken earlier or later than these had a survival rate of only 20–30%.

3. Nursery location. Bamboo cuttings do not store water so the nursery should be located in a place with easy access to irrigation, excellent drainage and a friable, fertile soil. Avoid water deficient locations with poorly drained, clayey soils.

4. Nursery management. During the nursery stage regular watering is vital, particularly in the period between the emergence of buds above soil level and the production of roots. At this time the propagules have a "falsely alive" appearance (they are shoots without roots) and if they are not supplied with enough water the newly formed shoots will die in large numbers. Fertilizer should be added once the shoots have produced roots.

2.2. PROPAGATION BY SECONDARY BRANCH CUTTINGS

Secondary branches are those that grow from the basal portion of the primary branches at each node. They have a large rhizome at the base with root initials and under suitable conditions can be enticed to grow into complete plants. The main advantages of secondary branch cuttings over other propagation methods are that the culm does not need to be cut down, transportation is easier, less labour is required and the cost of each propagule is low. Table 11 shows the results of experiments on secondary branch cuttings on 6 genera of bamboos in January 1991 in Guangdong Forestry Research Institute. Over half the species had a success rate of over 50%, which illustrates the widespread suitability of this method for the propagation of many species.

2.2.1. Propagation method with secondary branch cuttings

1. Take secondary branch cuttings in the spring, and cut off the tips leaving propagules about 30–40 cm long.

2. Place them in clean water in a cool and shady environment in preparation for planting.

3. Form irrigation channels on well-prepared nursery beds and bury the cuttings slanting at about 10–12 degrees to the horizontal, with the rhizome portion about 3–6 cm deep and the uppermost node just above soil level.

Table 11.
Propagation of bamboos by secondary branch cuttings (January 1991)

Species	Number of cuttings	Number of plants produced	Percentage survival
B. chungii	25	14	56.0
B. gibboides	20	10	50.0
B. pervariabilis	90	22	24.4
B. pervariabilis × (Dendrocalamus latiflorus + B. textilis)	180	109	60.6
B. textilis	120	72	60.8
B. tulda	27	9	33.3
B. tuldoides cv.	33	8	24.2
B. vario-striata	90	50	55.5
B. viridi-vitata	16	7	43.8
Bambusa vulgaris	90	86	95.5
Dendrocalamus bambusoides	16	2	12.5
Dendrocalamus brandisii	30	8	26.6
Dendrocalamus minor var amoenus	8	7	87.5
Dinchloa utilis	13	3	23.1
Gigantochloa levis	90	45	50.0
Neosinocalamus affinis	90	28	31.1
Thyrsostachys siamensis	5	4	80.0

4. Cover with soil, add mulch and water-in thoroughly.

Roots are formed in 7–15 days after planting and buds germinate in 20–30 days. The success rate on a large scale is about 60% and may reach over 90%.

2.2.2. Main factors influencing success of secondary branch cuttings

2.2.2.1. Age of branch. The success rate is highest with branches that are between six months and one year-old. The tissues of branches younger than this are still immature, they produce shoots very slowly and may rot if in the ground for too long. Older branches cannot produce roots themselves and such propagules will only form roots from the rhizomes of newly grown shoots. These "falsely alive" shoots and require a great deal of attentive nurturing.

Characteristics to look out for when selecting six month to one year-old branches for use in propagation are: stems dark green, sheaths fallen or almost completely fallen-off, a few leaves expanded on the tip of the branch and nodal buds not yet germinated or just starting to germinate.

2.2.2.2. Season. Cuttings should be taken in early spring before the buds on the branch have germinated or just as they are starting to germinate. Success rates are much lower if taken after germination is underway. For example in 1989 at GDFRI we trialled 0.5–0.8 cm diameter branches taken at different times in the spring. On

6 March (just at the time the buds were starting to germinate) the survival rate was 56.3%, but by 24 March the survival rate was only 12%.

2.2.2.3. Branch diameter.　The thicker the branch the greater the chance of success. At GDFRI we conducted trials with induced branches (see Section 2.2.3) and tried three groups: 3–5 mm diameter, 6–8 mm, and 8 mm and above, with success rates of 31.1%, 56.3% and 91.3% respectively.

2.2.2.4. Nursery management.　It is important to keep the cuttings well watered but avoid puddling on the soil surface because this will cause rotting of the roots. Fertilize regularly and lightly once the new shoots have emerged above the soil surface. For the first dose use 225 kg/ha urea and then fertilize every two weeks, gradually increasing the quantity of fertilizer as the plantlets grow. Note that adding nitrogen as part of a compound fertilizer is more beneficial to growth than applying it on its own, as compound fertilizers also increase the number of buds germinating and increase the fresh weight of the plants.

2.2.3. Secondary branch-propagule production nursery

Under natural conditions the number of secondary branches is limited. Many are located on the upper nodes of the culm and are very difficult to collect. The use of a secondary branch-propagule production nursery allows the production of a huge number of secondary branches which will all be approximately the same age. The method is as follows.

2.2.3.1. Selection of propagule.　Collect healthy one-year-old culms, 2 cm in diameter, from plants within the nursery, because these will have well-developed root systems and are more likely to survive.

2.2.3.2. Culm extraction.　Cut back the culm to a length of 1.5 metres in order to reduce apical dominance over the buds on the lower part of the culm and encourage them to germinate.

2.2.3.3. Damaging the rhizome buds.　Dig out the culm and remove the rhizome buds with a knife to prevent them from producing shoots (Fig. 10). This will ensure more nutrients are available for the germinating branches. Research indicates that this is a key stage in the process.

2.2.3.4. Planting and management.　Plant the propagules at 1 × 1 m or 1 × 1.5 m spacings in the nursery. Water, fertilize and improve the light regime in the nursery to produce an ideal environment for producing secondary branches.

Using this method 8–10 secondary branch cuttings can be obtained from each culm in the first year, and possibly up to 18 per culm. It is estimated that a

Figure 10. Removing the rhizome buds on secondary branch cuttings is vital for success.

Figure 11. Secondary branch cuttings in the propagation bed.

minimum of 53,000 branch cuttings could be produced per hectare by the spring of the following year. The cuttings are high quality, all of a similar age and the survival rate is high. In previous tests with one thousand cuttings the success rate was 60.60%. The use of this technique permits the mass propagation of a large number of propagules (Fig. 11) which reduces the manual input in the bamboo nursery and speeds the distribution of bamboo propagules.

2.3. PROPAGATION BY WHOLE CULM-CUTTINGS

1. Sever 1–2 year old culms from their rhizome and cut off the tip.
2. Remove all except one broad branch at each node and cut it back to just above the first node.
3. Cut a 2–5 cm diameter hole in each internode.
4. Plant the whole length of culm horizontally in the soil, cover with 6–10 cm of soil and mulch to maintain moisture. Culms should be spaced 20 cm apart.

This method ensures proper growth and rooting of nodal buds because nutrient transport to the base has been hindered by the removal of some of the conducting tissues between the nodes. The success rate is quite high. However on removal from the nursery the culm needs to be cut into individual plantlets which can be troublesome. This method should only be used for bamboos in which success rates are low with other methods, or for bamboos which have been transported over long distances (and over long periods of time) and are therefore not in optimum condition when planted.

2.4. PROPAGATION BY CULM DIVISIONS

Culm divisions can be used to produce yet more propagules from the young plants in the whole culm nursery, the nodal cuttings nursery and the secondary branches nursery. This method is not confined to particular seasons and clumps can be divided many times in one year. The practice is quite simple, the costs are low and the success rate is high. The steps are:

1. Select a well growing seedling clump in the nursery and cut off two-thirds of the leaves. Cut back the leafy branches to a length of 50–60 cm.
2. Dig out the soil around the roots and remove soil from the rhizome area in order to expose the join.
3. Gradually separate the culms using your hands and a pair of secateurs, being careful not to damage the rhizomes or the root systems.
4. Divide the cuttings into grades, take them to the nursery and plant healthy, undamaged cuttings at 15 × 30 cm spacings.

5. Damaged or old cuttings need to be buried. Plant at an angle with the base 10-cm deep and leave just one node sticking out the surface of the soil. Cover, firm-in, add mulch and water.

Water once or twice each day in the first week and fertilize when the culm has started producing new roots and leaves. Initially use nitrogen at 225–300 kg/ha of urea and subsequently fertilize every 2 weeks increasing the dosage as the plants grow.

2.5. RAPID MULTIPLICATION BY TISSUE CULTURE

Rapid propagation by tissue culture involves culturing plant organs, such as the embryo, apical buds, stem sections and spikelets (collectively known as "explants") on nutrient-rich media in sterile culture bottles and inducing them to proliferate by incorporating plant growth regulators in the medium. The main advantage of this technique over conventional seedling propagation methods is the huge number of propagules that can be produced in the same space of time. One single bud can be multiplied 1–5 times at each passage and within a year 10 000–100 000 or more propagules can be produced from that bud. This is the most efficient propagation method presently available.

Tissue culture requires the provision of specific facilities and conditions, and highly skilled workers. So, whilst promoting the use of tissue culture methods, conventional propagation methods should not be overlooked. Tissue culture can be used to enlarge the propagation base, and conventional propagation can be used on a more local scale. By combining both these techniques bamboos can be propagated rapidly and economically.

There are two methods of producing plants by tissue culture. The first is via callus. This involves firstly dedifferentiation of the explant into a mass of cells called callus, and then the differentiation of organs on the callus leading to the formation of whole plants. In this procedure the dedifferentiation stage requires the use of 2,4-dichlorophenoxyacetic acid. This is able to induce genetic mutations in cells and so this method is inappropriate for propagating plants in which the genetic characteristics of the selected hybrids must be maintained.

The second method is by rapid proliferation of buds. In this case the explant does not pass through a callus phase but is induced to produce new clumps of buds directly. The resulting shoots are then induced to produce more buds by transferring them to fresh media. This is a theoretically limitless procedure and is the preferred method for the rapid multiplication of new hybrids.

Tissue culture of bamboo hybrids includes the following stages: selection of the explant, sterilization of the explant, induction of buds, proliferation of buds, root formation, and transfer to pots. Two types of explant may be utilized; embryos (i.e. seeds) and stem portions. The techniques involved at each stage are discussed below.

2.5.1. Embryo culture

2.5.1.1. Sterilization of the explant. Seeds should be sterilized for 15–30 minutes with 0.1% mercuric chloride and sodium hypochlorite solution (2.5–5% chlorine content). They may be used alone or in combination but experiments have shown that using them together gives the best results. Sodium hypochlorite (NaOCl) damages soft tissues, so dry seeds should be first immersed in 2.5% NaOCl for 10 minutes, and then placed in 0.1% mercuric chloride for 15 minutes. In this way the surface of the seed is sterilized and damage to the young embryo by NaOCl is avoided. This is a relatively safe and reliable method of sterilization. Using this method the seed sterilization rate is about 93%.

2.5.1.2. Induction of adventitious buds. Sterilized seeds will germinate on Murashige and Skoogs (MS) basal medium containing 0.2 mg/l benzyladenine (BA). When the shoot is 1 cm long the seed can be transferred to MS medium containing 2–4 mg/l of BA for the induction of buds.

2.5.1.3. Continuous proliferation of buds. Using MS basal medium with BA at concentrations of 1–3 mg/l bud propagation and growth is promoted. At concentrations of BA lower than 1 mg/l the rate of bud proliferation is low, and at concentrations higher than 3 mg/l growth malformations can occur.

Whichever medium is being used, the bud producing capacity and growth of the clones will gradually decline with time. In order to limit this loss of vigor, one can adopt a method of alternately culturing shoots on media containing high concentrations of BA for bud proliferation, and then low concentrations of BA, to allow the shoots to grow. Using this method it is possible to guarantee a specific propagation rate and the shoots grow well. Moreover the vigor of the clones will be maintained for longer.

Table 12 shows the results of comparative investigations on the growth of clones of *D. latiflorus* number 17 after one year. The results show that using this method the multiplication rate of the clones had not fallen significantly after 12 subcultures.

2.5.1.4. Induction of roots. Half-strength basal MS medium with different concentrations of the growth regulators naphthalene acetic acid (NAA) and indole butyric acid (IBA) are used for root induction. NAA has been found to be more effective than IBA between 1–3 mg/l for *D. latiflorus* (Table 13). The rooting percentage increased with increasing concentration in the medium. However at concentrations of 3 mg/l NAA or IBA some plants died in culture and so concentrations below 3 mg/l are recommended. Mixing the two growth regulators was more effective than using one alone and plantlet mortality should not be a problem as long as the total concentration of growth regulator in the medium does not exceed 3 mg/l. Moreover the rooting rate exceeds 90%.

Table 12.
Proliferation of *D. latiflorus* shoots on different media combinations

Original medium	Medium transferred to	Culture period (1992)	Number of shoots transferred	Number of new shoots formed	Rate of multiplication
MS, $BA_{0.5}$, $NAA_{0.1}$	MS, $BA_{0.5}$, $NAA_{0.1}$	April 19–May 22	78	21	1.269
¾MS, BA_1, $KT_{0.5}$	MS, $BA_{0.5}$, $NAA_{0.1}$	August 1–22	136	149	2.095
MS, $BA_{0.5}$, $NAA_{0.1}$	MS, BA_2	August 8–22	177	222	2.254
MS, $BA_{0.5}$, $NAA_{0.1}$	¾MS, BA_1, $KT_{0.5}$	August 6–22	174	218	2.253
MS, $BA_{0.5}$, $NAA_{0.1}$	MS, BA_4	August 7–22	192	218	2.135

Table 13.
Root induction of *D. latiflorus* shoots in culture

Clone	Growth regulators	Concentration mg/l	Number of clumps innoculated	Number of rooted clumps	Percentage rooting
1	NAA	1	30	14	46.7
2	NAA	2	30	17	56.7
3	NAA	3	30	26	86.7
4	IBA	1	30	10	33.3
5	IBA	2	30	14	46.7
6	IBA	3	30	21	70.0
7	NAA + IBA	1 + 1	30	26	86.7
8	NAA + IBA	1.5 + 1.5	30	28	93.3

2.5.1.5. Transfer from culture to pots. Transplanting survival rates of over 90% can be achieved if the following steps are taken:

1. Always transplant small clumps of shoots, never single shoots.

2. Transplant when the temperature is between 20–30°C. In Guangdong this is from March to June and September to December. From July to September the temperature and humidity are too high and the transplants are easily infected.

3. Transplant into river sand which is an ideal substrate, sterilized with a 1/5000 solution of potassium permanganate.

4. Maintain a very humid environment around the plantlets in the first two weeks after transplanting to acclimatize them.

Plantlets will recover about a week after transfer to pots, and will start to produce new leaves and shoots after about 20–30 days. After about 45 days they can be transplanted into the nursery.

2.5.2. Stem culture

2.5.2.1. Explant selection. Generally semi-woody branches are selected. They can be recognized by looking for branches which have just finished growing and which have a few leaves on them, but on which the side buds have not yet, or have only just, started to grow. Success is almost guaranteed with these explants. If the branch is too young then it can easily be damaged during sterilization, and if it is too old it is not easy to cut, the tissues are quite hard and manipulations *in vitro* are difficult.

2.5.2.2. Sterilization of the explant. This is one of the most important stages in the procedure. Bamboo internodes are hollow and this causes particular problems when sterilizing. Replicated experiments are required to find the optimum sterilizing agent and length of time of sterilization for your particular hybrid and location.

The stages involved in sterilization are:

1. Cut off the leaves from the selected explants, and wash the surface with clean water.

2. Place explants in a refrigerator at 4°C for one or two days to promote bud germination.

3. Immerse in 75% ethanol for one minute on removal from the refrigerator.

4. On a sterile surface immerse in 2.5% NaOCl for 10–15 minutes and wash in three changes of sterile water.

5. Transfer to 0.2% mercuric chloride and wash finally in 5 changes of sterile water.

2.5.2.3. Induction of buds. After sterilization remove the sheath and cut back the two ends to leave a single-noded explant with about 1 cm above and 1 cm below the node. This can then be planted in the medium. Use MS, 3/4 MS or 1/2 MS with BA at 3–6 mg/l.

2.5.2.4. Proliferation and growth of buds. The concentration of BA in the proliferation medium should be lower than in the induction medium to prevent the buds proliferating too rapidly and loosing some of their vigor. MS or 3/4 MS is still used as the basal medium. Alternating between media containing high and low concentrations of BA (as described in Section 2.5.1.3) helps maintain vigor and quality.

2.5.2.5. Rooting of shoot clusters. Half strength MS is usually used in conjunction with a suitable rooting initiator and a lower concentration of sucrose than for shoot growth. Combinations of NAA and IBA (as described in Section 2.5.1.4) give optimum rooting results. Shoots should be transferred to the rooting medium when they are about 1 cm long. The rooting percentage can be as high as 96%.

2.5.2.6. Transfer to pots. After 20–30 days when the roots have just started to grow the plantlets can be transferred to pots filled with sand sterilized with potassium permanganate solution for acclimatization. As with embryo-derived plantlets the optimum time to do this in Guangzhou is from the end of March to the beginning of June, since when the temperature exceeds 30°C the plantlets are less likely to survive. Cover completely with a thin, clear plastic film for the first two weeks in order to maintain humidity, and later pull back the edges in the early morning to allow air movement and gradually remove the cover altogether. Timely removal of diseased plants is required as is spraying with carbendazim at regular intervals. Mean survival rates of about 70% are achievable with these steps, and may be as high as 90% or more.

Plantlets recover their vigor after about a week in the sand, produce new leaves after 20–30 days and can be transplanted to the nursery after about 45 days.

2.5.3. Other techniques relevant to both embryo and shoot culture

2.5.3.1. Culture rejuvenation. After many passages the vigor of the cultures gradually starts to decline, the plantlets become small and weak and the bud proliferation rate falls. In order to maintain a fixed rate of propagation and healthy plants, it is necessary to rejuvenate the culture system by collecting fresh explants from plantlets that have already been produced by tissue culture and that are growing in the nursery and reculturing them. Individual lines should be eliminated after about 20 culture passages.

2.5.3.2. Multi-clone propagation. Mass propagation of a single clone is a risky strategy since it may suffer from mass flowering or lack resistance to one of many pests or diseases and cause huge losses when in the field. To prevent this by maintaining genetic diversity, a range of elite clones should be selected for rapid propagation.

Aside from the key steps outlined above, the success rate of micropropagation is related to the age of the explant (from seed germination all the way to harvest). The younger the explant, the greater the chance of success, and tissue culture becomes progressively more difficult as the explants get older. Tissue culture of mature bamboos is more difficult than of embryos, seedlings or young plants. Buds are difficult to induce, growth is poor and many die at the rooting stage. At GDFRI we succeeded in culturing 27 year-old plants of hybrid ChengMa 7 using BA and kinetin (KT) but the multiplication rate and growth of plantlets could not compare with that of young bamboos.

Particular requirements for the tissue culture of old bamboos are:

1) Old bamboos have lower nutrient requirements for most elements than younger bamboos so half or three-quarter strength basal MS is suitable.

2) BA is used at 3–6 mg/l to initiate buds, but during the rapid proliferation stage BA concentrations should be kept below 3 mg/l. If the concentration is too high extension growth of the shoots will be reduced and non-germinating buds may be formed.

3) Incorporation of 10–15% coconut milk into the medium is beneficial for growth.

2.5.4. Use of tissue culture techniques in the early selection of seedling hybrids

Bamboos are rapid growing and high-yielding plants but require a large area of land. Only 400–900 clumps can be grown per hectare. Evaluating seedling hybrids in the nursery also involves a large area of land and a relatively heavy workload. On comparing the growth of bamboos in culture and in the nursery we found there was almost no difference between performance in culture and in the nursery (e.g. productivity, culm shape, branching habits, leaf colour, shooting ability, adaptability). For example if plants grew rapidly in culture with deep green leaves and strong or moderate shoot-producing abilities then, when transplanted to the nursery the majority also grew vigorously and were relatively productive: if plants in culture were weak with light green leaves, then the majority of them did not grow well on transplanting to the nursery, then on transplanting to the nursery the majority were straight with few branches, and so on.

Thus it is possible to conduct early stage preliminary selection whilst the plants are still in culture in order to eliminate the very poor hybrids, leaving fewer, better clones for the next step of evaluation in the nursery. This reduces the amount of evaluation work required in the field.

According to research on the selection of seedling *D. latiflorus* bamboo clones conducted here at GDFRI, we have found that that tissue culture techniques can be effectively combined with hybridization and breeding work thus:

1) Inoculate seeds into culture after surface sterilization.

2) Propagate each of the clones separately so one can clearly see all the propagules of one clone in one location.

3) Remove the poor clones after six months.

4) Mass propagate the elite clones.

5) On achieving a predetermined number of propagules of each clone, take a portion of the cultures and root them.

6) Keep the other portion in culture for continued propagation or for germplasm storage.

7) Plant rooted clones in the nursery in standardized blocks to evaluate their growth and fitness.

8) Concurrently carry out evaluations of timber, pulping, shoot quality and anatomical investigations.

9) Identify elite clones for mass propagation on the basis of these investigations.

10) Mass propagate the elite clones from the proportion remaining in culture.

This procedure is a highly efficient method of selection and will shorten the time it takes from selection of a new hybrid to cultivation and use.

3   Induction of flowering

There are three reasons for attempting to induce flowering in bamboos.
1) Bamboos rarely flower and seed. This is the biggest obstacle to breeding. If it is possible to induce flowering then it may be possible to breed bamboos by the conventional methods as used for other crops and trees. With genetic improvements the economic value of bamboos will be even higher.
2) To supply seeds for afforestation.
3) To develop a greater understanding of the mechanism of flowering in bamboos, which would facilitate control of flowering and reduce the economic losses caused by the natural flowering of bamboo forests.

   In the last ten years there has been some progress in the induction of flowering in bamboos. Nadgauda *et al.* (1990) reported flowering of very young plants of *B. arundinacea* and *D. brandisii* in liquid media and the production of seeds. At GDFRI we noted seed-derived, four month-old *D. latiflorus* plantlets flowering in culture and subsequently began investigations into *in vitro* flowering. Some of the methods used to induce flowering in bamboos in China are outlined below.

3.1. FORCING BAMBOOS TO FLOWER

The main principle of this method is to halt vegetative growth under adverse circumstances and force the bamboo into reproductive growth. This has been used in Yunnan province to induce *D. affinis* into flowering and seeding. The method is as follows:
1) Remove all the leaves from the clump.
2) Dig out all the soil from around the rhizome portion.
3) Burn the rhizomes by lighting a fire of dry, waste organic matter on top of them.
4) Recover the rhizosomes with soil when flowers appear on the upper branches of the culm replace the soil over the rhizomes.

   Other workers in different locations have applied this method and in some cases flowers were formed and in other cases flowers were not formed. It is necessary to find out more about the effect of age of the chosen plant, time of application and strength of application before it will be possible to achieve reasonably reliable results.

3.2. *IN VITRO* FLOWERING

During rapid mass propagation by tissue culture, plantlets that suddenly produce extremely dense clumps of buds and shoots are likely to start flowering. Shoots formed from these highly vigorous bud-clumps soon cease extension growth and a swelling forms at the apex which indicates the flower buds are forming and that flowering will soon occur. Many different clones of *D. latiflorus* have flowered continuously at GDFRI and all flowered within six months to three years of initiating the cultures. Number 17 has flowered continuously and produced a large number of flower buds, and our experiences can be applied to the flowering of other bamboos. The main points are outlined below:

1) Continuous culture of *D. latiflorus* seedlings on high concentrations of BA (3–6 mg/l) leads to the formation of flowers within six months to three years. The higher the concentration of BA the faster the induction of flowering.

2) Continuous culture of fresh, sterilized flowers of *D. latiflorus* on a bud prolifer- ation medium induces them to produce new flowers and spikelets. This method increases the numbers of flower buds but the flowers are of poor quality, are somewhat abnormal and cannot be used as maternal parents.

3) Reducing the BA content in the medium or placing buds on a medium contain- ing KT (0.5–1.5 mg/l), NAA (0.25–1.5 mg/l) and coconut milk (10–15%) im- proves the quality of the flowers significantly. However successively produced flowers appear to shift gradually from reproductive to vegetative growth and the extent of this is related to KT content. When the KT content is low the branches merely extend, but when it is high vegetative shoots appear from the base of the flower cluster.

4) The proliferating ability of particular clones can act as a guide to whether it will be possible to induce flowering in that clone. Flowering can be readily induced in clones with strong proliferating ability but those with poor proliferating abilities are much more difficult.

The above are merely the results of preliminary research. Inducing flowering in bamboos involves many scientific difficulties in the fields of plant physiology, biochemistry, genetics and breeding. There is a considerable distance between present research and practical application but we have already seen that man can influence bamboo flowering to some extent. This offers hope that the biggest difficulty in bamboo breeding can be solved, and once that is achieved the door will be open for many breakthroughs in bamboo breeding.

4   The prospects for bamboo breeding

Because flowering of bamboos cannot yet be controlled, genetic improvement of bamboos in the near future will rely on the standard breeding methods outlined in this manual. Sporadically-flowering bamboos will be collected for hybridization, but there should be a clear breeding objective, based on market requirements or prospects for future use. The main breeding objectives are likely to be:

1) Breeding shoot-use bamboos with high nutritional contents and good flavours.

2) Breeding durable, drought and cold resistant bamboos for papermaking.

3) Breeding ornamental bamboos.

The greater use of the elite strains that are already available should also be considered. There are elite cultivated forms, cultivated varieties and clones within our present-day bamboo forests due to the long period of natural and human selection to which they have been subjected. Many of these elite germplasm resources are limited to just one specific location and are cultivated on only a small area. By digging them up, propagating them and planting them in new locations, genetic improvement is achieved in a very practical way. But in order to achieve this further research is required on the tissue culture of mature bamboos. Presently the instances of succesful tissue culture of older plants are limited, and the main problems are:

1) Difficulties in inducing bud growth.

2) Poor growth of the shoots once bud growth has been induced.

3) High mortality on transfer to rooting medium.

These are all related to the slower reactions of the cells as they loose vitality when they mature. Further work is required on the selection of basal media and cultural conditions for these older bamboos.

Biotechnology is very significant in bamboo breeding, and its use is likely to be far from limited to rapid propagation by tissue culture. By combining biotechnology with conventional propagation techniques it should be possible to reduce the time required for breeding a new hybrid.

Bibliography

1. Bamboo Research Centre and Nanjing College of Forest Products (1974). *The Cultivation of Bamboos.* Agriculture Press, Xian, 278 pp.
2. Chen FuShu (1986). Fibre analysis and selection of clump-forming bamboos for use in papermaking, *Guangdong Forestry Science* (2), 1-9.
3. Chen RuiYang and Song WenQin (1982). A new method for preparing chromosomes — The hypotonic method and it's use in cytology. In: *Plant chromosomes and chromosome techniques.* Science Press, Beijing, 204 pp.
4. Forest Management Research Centre and Guangdong Forestry Research Institute (1973). Studies on the biological properties of flowering and fruiting in *Bambusa pervariabilis* and *Bambusa textilis, Guangdong Forestry Science Reports* (3), 1-5.
5. Forest Management Research Centre and Guangdong Forestry Research Institute (1976). Initial studies on Hybridization of Bamboos, *Scientia Silvae Sinica* (2), 47-52.
6. Forest Management Research Centre and Guangdong Forestry Research Institute (1979). Further studies on hybridization of bamboos, *China Forest Science Reports* 1.
7. Guangdong Forestry Research Institute (1975). Studies on the determination and preservation of pollen viability in bamboos, *Guangdong Forestry Science Reports* (3), 2-5.
8. Heuch, J. H. R. (1992). In vitro flowering of bamboo species — Prospects and Aims. In: Zhu ShiLin, Li Weidong, Zhang XinPing and Wang ZhongMing (Eds). *Bamboo and its Use.* Proceedings of the International Symposium on Industrial Use of Bamboo, Beijing, 1992. International Tropical Timber Organization and Chinese Academy of Forestry, pp. 47-55.
9. Li ZhengLi (1973). *Botanical microtechniques.* Science Press, Beijing, 106 pp.
10. Nadgauda, R. S., Parsharami, V. A. and Mascarenhas, A. F. (1990). Precocious flowering and seeding behaviour in tissue-cultured bamboo, *Nature* **344**, 335-336.
11. Nadgir, A. L., Phadke, C. H., Gupta, P. K., Parasharam, V. A., Nair, S. and Mascarenhas, A. F. (1984). Rapid multiplication of bamboo by tissue culture, *Silvae Genetica* **33**, 219-223.
12. Nanjing College of Forestry Products (1980). *Tree genetics and breeding.* Science Press, Beijing, 457 pp.
13. Tan YuanJie (1994). Preliminary studies on *Dendrocalamus brandis* stem culture in vitro, *Journal of Bamboo Research* **13** (3), 16-21.
14. Varmah, J. C. and Bahadur, K. N. (1980). Country report and status of Research on bamboos in India, *Indian Forest Records* **1**, 1-28.
15. Wang YuXia and Zhang GuangChu (2000). Micropropagation and growth of *Dendrocalamus latiflorus* clones from seed, *Guangdong Forestry Science* **16** (3), 1-5.
16. Wu YiMing (1999). Current research on bamboo tissue culture and plantlet regeneration, *Journal of Bamboo Research* **18** (1), 32-37.
17. Yeh, M. L. and Chang, W. C. (1987). Plant regeneration via somatic embryogenesis in mature embryo-derived callus of *Sinocalamus latiflora, Plant Science* **51** (1), 93-96.
18. Zhang ChunXia, Xie YinFeng, Zhang YouFa, He DeTing, Chen TianBao and Wu WeiWen (1999). Progress and prospects in bamboo tissue culture, *Journal of Bamboo Research* **18** (3), 46-49.
19. Zhang GuangChu, Chen FuShu, Yang AiGuo and Fu MaoYi (1994). Propagation of clump-forming bamboos by secondary branch cuttings, *Journal of Southwest Forestry College* **14** (3), 137-142.

20. Zhang GuangChu, Chen FuShu and Yang AiGuo (1993). Studies on the selection of top-quality shoot producing bamboos, *Guangdong Forestry Science* (2), 9–13.
21. Zhang GuangChu and Wang YuXia (1998). Bamboo breeding: current status and future prospects, *Journal of Bamboo Research* **17** (1), 6–9.
22. Zhang GuangChu and Wang YuXia (2000). Preliminary studies on the *in vitro* flowering of bamboos, *Journal of Bamboo Research* **19** (4) (in press).
23. Zhang GuangChu and Chen FuShu (1986). Studies on Bamboo Hybridization, *Bamboo Research* (3), 48–53.
24. Zhang GuangChu and Chen FuShu (1983). ChengMaQing 1 — An excellent hybrid bamboo, *Scientia Silvae Sinica* **16** (Suppl.), 124–126.
25. Zhang GuangChu and Chen FuShu (1983). Early evaluation of culm quality in hybrid bamboos, *Bamboo Research* (2), 68–74.
26. Zhang GuangChu (1985). Studies on the chromosome numbers of some sympodial bamboos, *Bamboo Research* (2), 1–7.
27. Zhang GuangChu, Chen FuShu and Wang YuXia (1993). Studies on the rapid propagation *in vitro* of *Dendrocalamus latiflorus*, *Journal of Bamboo Research* **12** (4), 7–15.
28. Zhu HuiFang and Yao XiShen (1964). Fibre structure of 33 pulp-use bamboo species in China, *Forestry Science* **9** (4), 311–331.

Printed in the United States
by Baker & Taylor Publisher Services